WEIRD ANIMALS

AXOLOTL

AMY CULLIFORD

A Crabtree Roots Book

Crabtree Publishing

crabtreebooks.com

School-to-Home Support for Caregivers and Teachers

This book helps children grow by letting them practice reading. Here are a few guiding questions to help the reader with building his or her comprehension skills. Possible answers appear here in red.

Before Reading:

• What do I think this book is about?
  • *I think this book is about a weird animal called an axolotl.*
  • *I think this book is about where axolotls live.*

• What do I want to learn about this topic?
  • *I want to learn about what axolotls eat.*
  • *I want to learn why axolotls have weird things on their heads.*

During Reading:

• I wonder why...
  • *I wonder why axolotls have gills.*
  • *I wonder why axolotls live only in water.*

• What have I learned so far?
  • *I have learned that some axolotls can lay eggs.*
  • *I have learned that all axolotls grow legs.*

After Reading:

• What details did I learn about this topic?
  • *I have learned that axolotls have little teeth.*
  • *I have learned that axolotls swim in water.*

• Read the book again and look for the vocabulary words.
  • *I see the word **gills** on page 6 and the word **eggs** on page 10. The other vocabulary words are found on page 14.*

Splash! I see
an **axolotl**!

Axolotls live only
in water.

All axolotls have **gills**.

Axolotls use their little **teeth** to eat.

Some axolotls can lay **eggs**.

All axolotls
grow **legs**.

Word List
Sight Words

all	I	see
an	in	some
can	lay	their
eat	little	to
grow	live	use
have	only	water

Words to Know

axolotl

eggs

gills

legs

teeth

30 Words

Splash! I see an **axolotl**!

Axolotls live only in water.

All axolotls have **gills**.

Axolotls use their little **teeth** to eat.

Some axolotls can lay **eggs**.

All axolotls grow **legs**.

WEIRD ANIMALS
AXOLOTL

Written by: Amy Culliford

Designed by: Rhea Wallace

Series Development: James Earley

Proofreader: Melissa Boyce

Educational Consultant: Marie Lemke M.Ed.

Photographs:
Shutterstock: Natalay Studio: cover; Tum UR: p. 1; Maria
  Dry Shout: p. 3; Nyura: p. 5; Kostyantyn Sulima: p. 7;
  MK Studio: p. 9; fotocat5: p. 11; Lord Benedict: p. 13

Crabtree Publishing

crabtreebooks.com   800-387-7650

Printed in Canada/042024/CPC20240409

Published in Canada
Crabtree Publishing
616 Welland Ave.
St. Catharines, Ontario
L2M 5V6

Published in the United States
Crabtree Publishing
347 Fifth Ave
Suite 1402-145
New York, NY 10016

Library and Archives Canada Cataloguing in Publication
Available at Library and Archives Canada

Library of Congress Cataloging-in-Publication Data
Available at the Library of Congress

Hardcover: 978-1-0398-0983-3
Paperback: 978-1-0398-1036-5
Ebook (pdf): 978-1-0398-1142-3
Epub: 978-1-0398-1089-1

Word Count: 30
GRL: D
Subject: Animals
Genre: Science
Standard: K-ESS3-1
Vocabulary Words: axolotl, eggs, gills, legs, teeth

AXOLOTL

This book introduces early readers to the axolotl. Simple text and vibrant images help engage children and grow a love of reading!

Books in *Weird Animals* include:

Axolotl   *Frill-necked Lizard*   *Pangolin*

Aye-aye   *Hairy Frog*   *Platypus*

Crabtree Roots

These books feature simple text with each picture to aid comprehension and help children learn to read with confidence. Every book begins with a list of questions to help caregivers and teachers guide young readers.

Teacher's Guides

ISBN 978-1-0398-1036-5

90000

9 781039 810365

✤ Crabtree Publishing
crabtreebooks.com